MATH
WORKBOOK

ADDITION & SUBTRACTION

ADDITION

1)   142
  + 697
  ....

2)   375
  + 373
  ....

3)   806
  + 540
  ....

4)   471
  + 115
  ....

5)   456
  + 696
  ....

6)   549
  + 242
  ....

7)   138
  + 907
  ....

8)   205
  + 537
  ....

9)   734
  + 464
  ....

10) 960
  + 289
  ....

11) 463
  + 243
  ....

12) 209
  + 263
  ....

13) 650
  + 758
  ....

14) 494
  + 438
  ....

15) 602
  + 256
  ....

16) 839
  + 453
  ....

17) 583
  + 829
  ....

18) 784
  + 462
  ....

19) 327
  + 610
  ....

20) 223
  + 291
  ....

(1)   146
    + 133
....

(2)   511
    + 300
....

(3)   856
    + 196
....

(4)   658
    + 309
....

(5)   223
    + 742
....

(6)   309
    + 223
....

(7)   340
    + 769
....

(8)   369
    + 712
....

(9)   401
    + 607
....

(10)   116
    + 491
....

(11)   236
    + 936
....

(12)   619
    + 778
....

(13)   934
    + 153
....

(14)   798
    + 244
....

(15)   491
    + 885
....

(16)   874
    + 215
....

(17)   887
    + 175
....

(18)   246
    + 885
....

(19)   339
    + 641
....

(20)   744
    + 151
....

1) 397
  + 514

2) 817
  + 336

3) 867
  + 194

4) 335
  + 265

5) 623
  + 483

6) 473
  + 759

7) 862
  + 964

8) 891
  + 821

9) 602
  + 543

10) 265
   + 837

11) 607
   + 913

12) 175
   + 764

13) 388
   + 893

14) 119
   + 657

15) 263
   + 137

16) 472
   + 606

17) 377
   + 940

18) 753
   + 186

19) 596
   + 198

20) 308
   + 519

①	176 + 677	②	102 + 586	③	111 + 866	④	482 + 921

⑤	311 + 805	⑥	325 + 504	⑦	335 + 744	⑧	975 + 326

⑨	652 + 488	⑩	511 + 550	⑪	776 + 886	⑫	318 + 173

⑬	531 + 501	⑭	138 + 850	⑮	743 + 369	⑯	746 + 542

⑰	250 + 334	⑱	444 + 284	⑲	378 + 774	⑳	675 + 486

1.　　828
　　+ 631

2.　　278
　　+ 941

3.　　159
　　+ 782

4.　　355
　　+ 100

5.　　401
　　+ 791

6.　　947
　　+ 877

7.　　812
　　+ 125

8.　　195
　　+ 581

9.　　965
　　+ 684

10.　　308
　　+ 431

11.　　765
　　+ 315

12.　　399
　　+ 772

13.　　537
　　+ 794

14.　　900
　　+ 752

15.　　137
　　+ 255

16.　　824
　　+ 449

17.　　719
　　+ 165

18.　　656
　　+ 823

19.　　117
　　+ 978

20.　　573
　　+ 182

① 881
+ 601

② 962
+ 246

③ 965
+ 597

④ 258
+ 747

⑤ 468
+ 372

⑥ 564
+ 985

⑦ 211
+ 226

⑧ 585
+ 131

⑨ 792
+ 639

⑩ 749
+ 583

⑪ 656
+ 385

⑫ 451
+ 421

⑬ 931
+ 151

⑭ 728
+ 995

⑮ 321
+ 143

⑯ 590
+ 611

⑰ 574
+ 309

⑱ 300
+ 344

⑲ 771
+ 376

⑳ 793
+ 525

1. $868 + 262$

2. $526 + 425$

3. $780 + 952$

4. $511 + 901$

5. $925 + 341$

6. $358 + 516$

7. $846 + 325$

8. $788 + 265$

9. $237 + 422$

10. $269 + 694$

11. $487 + 234$

12. $857 + 507$

13. $321 + 897$

14. $855 + 654$

15. $600 + 703$

16. $193 + 230$

17. $163 + 877$

18. $655 + 633$

19. $963 + 821$

20. $348 + 298$

① 197
+ 856
....

② 401
+ 322
....

③ 248
+ 228
....

④ 453
+ 967
....

⑤ 669
+ 255
....

⑥ 301
+ 415
....

⑦ 174
+ 970
....

⑧ 869
+ 725
....

⑨ 564
+ 960
....

⑩ 952
+ 915
....

⑪ 152
+ 425
....

⑫ 522
+ 158
....

⑬ 343
+ 365
....

⑭ 757
+ 544
....

⑮ 449
+ 920
....

⑯ 329
+ 945
....

⑰ 717
+ 465
....

⑱ 201
+ 968
....

⑲ 460
+ 446
....

⑳ 748
+ 426
....

(1)　470
　 + 909

(2)　618
　 + 748

(3)　925
　 + 356

(4)　697
　 + 708

(5)　869
　 + 900

(6)　365
　 + 218

(7)　313
　 + 626

(8)　964
　 + 274

(9)　564
　 + 546

(10)　816
　 + 299

(11)　611
　 + 356

(12)　794
　 + 647

(13)　749
　 + 885

(14)　879
　 + 530

(15)　634
　 + 429

(16)　634
　 + 218

(17)　467
　 + 865

(18)　956
　 + 142

(19)　359
　 + 212

(20)　151
　 + 984

① 476   ② 174   ③ 366   ④ 750
  + 267   + 540   + 247   + 447

⑤ 669   ⑥ 914   ⑦ 495   ⑧ 851
  + 922   + 452   + 831   + 566

⑨ 724   ⑩ 209   ⑪ 755   ⑫ 307
  + 946   + 697   + 440   + 977

⑬ 222   ⑭ 692   ⑮ 594   ⑯ 927
  + 746   + 688   + 842   + 388

⑰ 213   ⑱ 609   ⑲ 970   ⑳ 581
  + 110   + 280   + 191   + 391

①   554   ②   852   ③   994   ④   340
  + 162   + 826   + 625   + 485
  ....

⑤   322   ⑥   594   ⑦   464   ⑧   395
  + 671   + 633   + 818   + 972
  ....

⑨   570   ⑩   335   ⑪   354   ⑫   760
  + 905   + 368   + 998   + 789
  ....

⑬   454   ⑭   526   ⑮   725   ⑯   319
  + 759   + 810   + 573   + 875
  ....

⑰   773   ⑱   836   ⑲   930   ⑳   406
  + 962   + 339   + 143   + 216
  ....

1) 199
  + 528

2) 759
  + 748

3) 452
  + 719

4) 859
  + 502

5) 981
  + 156

6) 142
  + 526

7) 599
  + 219

8) 754
  + 347

9) 627
  + 505

10) 580
  + 598

11) 956
  + 483

12) 804
  + 413

13) 108
  + 265

14) 115
  + 724

15) 697
  + 381

16) 937
  + 867

17) 670
  + 171

18) 117
  + 185

19) 347
  + 721

20) 902
  + 590

1. 704
    + 235

2. 804
    + 700

3. 219
    + 274

4. 855
    + 528

5. 202
    + 668

6. 100
    + 904

7. 602
    + 989

8. 602
    + 656

9. 422
    + 855

10. 143
    + 968

11. 710
    + 374

12. 430
    + 680

13. 453
    + 141

14. 385
    + 796

15. 792
    + 401

16. 461
    + 900

17. 980
    + 715

18. 885
    + 860

19. 596
    + 460

20. 830
    + 196

① 888
  + 159

② 735
  + 801

③ 307
  + 199

④ 924
  + 886

⑤ 204
  + 741

⑥ 160
  + 961

⑦ 124
  + 947

⑧ 786
  + 685

⑨ 736
  + 703

⑩ 663
  + 430

⑪ 747
  + 792

⑫ 525
  + 171

⑬ 108
  + 593

⑭ 598
  + 308

⑮ 238
  + 556

⑯ 156
  + 472

⑰ 213
  + 753

⑱ 307
  + 223

⑲ 389
  + 836

⑳ 519
  + 917

① 600
+ 716

② 748
+ 647

③ 685
+ 688

④ 847
+ 371

⑤ 740
+ 662

⑥ 395
+ 899

⑦ 285
+ 744

⑧ 333
+ 341

⑨ 405
+ 895

⑩ 510
+ 675

⑪ 749
+ 243

⑫ 322
+ 347

⑬ 862
+ 954

⑭ 910
+ 297

⑮ 423
+ 212

⑯ 590
+ 724

⑰ 143
+ 805

⑱ 417
+ 481

⑲ 757
+ 840

⑳ 943
+ 972

① 777
 + 651

② 946
 + 628

③ 173
 + 670

④ 856
 + 479

⑤ 428
 + 851

⑥ 583
 + 564

⑦ 951
 + 537

⑧ 192
 + 277

⑨ 528
 + 898

⑩ 750
 + 170

⑪ 139
 + 195

⑫ 612
 + 335

⑬ 421
 + 852

⑭ 570
 + 925

⑮ 842
 + 958

⑯ 985
 + 356

⑰ 156
 + 314

⑱ 296
 + 931

⑲ 625
 + 185

⑳ 350
 + 924

① 200
+ 930

② 636
+ 469

③ 865
+ 327

④ 508
+ 737

⑤ 341
+ 260

⑥ 651
+ 443

⑦ 648
+ 825

⑧ 233
+ 602

⑨ 707
+ 780

⑩ 729
+ 195

⑪ 168
+ 853

⑫ 900
+ 157

⑬ 130
+ 489

⑭ 605
+ 441

⑮ 623
+ 613

⑯ 893
+ 913

⑰ 877
+ 131

⑱ 825
+ 106

⑲ 899
+ 118

⑳ 965
+ 378

① 720
\+ 168

② 213
\+ 727

③ 366
\+ 738

④ 564
\+ 617

⑤ 100
\+ 227

⑥ 701
\+ 689

⑦ 814
\+ 450

⑧ 378
\+ 121

⑨ 477
\+ 136

⑩ 922
\+ 543

⑪ 541
\+ 681

⑫ 859
\+ 535

⑬ 639
\+ 636

⑭ 930
\+ 765

⑮ 878
\+ 579

⑯ 975
\+ 420

⑰ 678
\+ 589

⑱ 657
\+ 749

⑲ 933
\+ 748

⑳ 598
\+ 495

1.　　807
　 + 798

2.　　738
　 + 520

3.　　251
　 + 381

4.　　111
　 + 325

5.　　793
　 + 629

6.　　187
　 + 728

7.　　537
　 + 527

8.　　522
　 + 211

9.　　756
　 + 505

10.　　149
　 + 354

11.　　194
　 + 399

12.　　932
　 + 747

13.　　710
　 + 982

14.　　763
　 + 782

15.　　655
　 + 617

16.　　541
　 + 356

17.　　660
　 + 868

18.　　688
　 + 322

19.　　691
　 + 504

20.　　303
　 + 205

①	618 + 328	②	669 + 276	③	188 + 830	④	168 + 668
⑤	838 + 300	⑥	520 + 352	⑦	302 + 944	⑧	980 + 876
⑨	560 + 728	⑩	703 + 571	⑪	527 + 633	⑫	408 + 832
⑬	334 + 738	⑭	727 + 826	⑮	334 + 280	⑯	456 + 700
⑰	694 + 136	⑱	586 + 145	⑲	648 + 662	⑳	216 + 314

SUBTRACTION

① 952
  - 833

② 578
  - 104

③ 276
  - 160

④ 725
  - 153

⑤ 640
  - 575

⑥ 606
  - 458

⑦ 630
  - 132

⑧ 103
  - 103

⑨ 478
  - 202

⑩ 403
  - 241

⑪ 418
  - 299

⑫ 619
  - 154

⑬ 612
  - 191

⑭ 759
  - 311

⑮ 447
  - 304

⑯ 303
  - 218

⑰ 138
  - 116

⑱ 867
  - 137

⑲ 575
  - 432

⑳ 917
  - 249

① 645
− 305
....

② 728
− 107
....

③ 688
− 678
....

④ 698
− 194
....

⑤ 286
− 278
....

⑥ 459
− 359
....

⑦ 426
− 150
....

⑧ 661
− 381
....

⑨ 472
− 443
....

⑩ 163
− 151
....

⑪ 492
− 248
....

⑫ 798
− 425
....

⑬ 619
− 534
....

⑭ 680
− 538
....

⑮ 108
− 104
....

⑯ 788
− 169
....

⑰ 333
− 308
....

⑱ 619
− 341
....

⑲ 238
− 128
....

⑳ 101
− 100
....

1) $404 - 377$

2) $233 - 190$

3) $272 - 151$

4) $517 - 297$

5) $619 - 183$

6) $413 - 167$

7) $929 - 117$

8) $749 - 652$

9) $305 - 139$

10) $950 - 345$

11) $825 - 726$

12) $183 - 107$

13) $663 - 328$

14) $766 - 522$

15) $377 - 196$

16) $878 - 817$

17) $790 - 226$

18) $741 - 216$

19) $493 - 286$

20) $456 - 179$

① 602
  - 517
  ....

② 247
  - 134
  ....

③ 364
  - 213
  ....

④ 501
  - 397
  ....

⑤ 377
  - 250
  ....

⑥ 471
  - 380
  ....

⑦ 813
  - 616
  ....

⑧ 516
  - 204
  ....

⑨ 693
  - 449
  ....

⑩ 559
  - 528
  ....

⑪ 970
  - 450
  ....

⑫ 756
  - 478
  ....

⑬ 412
  - 242
  ....

⑭ 670
  - 454
  ....

⑮ 253
  - 136
  ....

⑯ 426
  - 222
  ....

⑰ 482
  - 333
  ....

⑱ 696
  - 694
  ....

⑲ 454
  - 445
  ....

⑳ 288
  - 128
  ....

① 813
  - 687

② 613
  - 135

③ 261
  - 106

④ 397
  - 154

⑤ 154
  - 125

⑥ 869
  - 626

⑦ 441
  - 223

⑧ 715
  - 267

⑨ 433
  - 222

⑩ 331
  - 123

⑪ 854
  - 153

⑫ 944
  - 444

⑬ 481
  - 150

⑭ 840
  - 328

⑮ 813
  - 461

⑯ 905
  - 749

⑰ 185
  - 132

⑱ 420
  - 236

⑲ 214
  - 210

⑳ 265
  - 146

① 294
- 104
....

② 450
- 294
....

③ 614
- 126
....

④ 630
- 104
....

⑤ 105
- 101
....

⑥ 230
- 153
....

⑦ 264
- 260
....

⑧ 268
- 168
....

⑨ 609
- 468
....

⑩ 550
- 232
....

⑪ 354
- 261
....

⑫ 888
- 294
....

⑬ 748
- 204
....

⑭ 526
- 234
....

⑮ 435
- 312
....

⑯ 584
- 566
....

⑰ 385
- 139
....

⑱ 158
- 152
....

⑲ 141
- 139
....

⑳ 679
- 596
....

① 617
- 242
....

② 671
- 313
....

③ 749
- 218
....

④ 182
- 104
....

⑤ 919
- 340
....

⑥ 422
- 298
....

⑦ 171
- 146
....

⑧ 775
- 397
....

⑨ 503
- 384
....

⑩ 859
- 361
....

⑪ 647
- 165
....

⑫ 749
- 604
....

⑬ 528
- 261
....

⑭ 370
- 360
....

⑮ 435
- 261
....

⑯ 772
- 706
....

⑰ 207
- 125
....

⑱ 354
- 168
....

⑲ 286
- 251
....

⑳ 504
- 189
....

1) 893
  - 201

2) 704
  - 360

3) 590
  - 324

4) 546
  - 477

5) 424
  - 256

6) 310
  - 234

7) 331
  - 244

8) 152
  - 105

9) 706
  - 355

10) 586
  - 553

11) 363
  - 174

12) 425
  - 116

13) 411
  - 196

14) 642
  - 441

15) 487
  - 419

16) 743
  - 324

17) 922
  - 185

18) 404
  - 148

19) 851
  - 111

20) 779
  - 629

① 868
- 765
....

② 474
- 300
....

③ 790
- 710
....

④ 750
- 199
....

⑤ 357
- 331
....

⑥ 339
- 258
....

⑦ 699
- 556
....

⑧ 896
- 144
....

⑨ 657
- 234
....

⑩ 961
- 829
....

⑪ 212
- 204
....

⑫ 278
- 230
....

⑬ 138
- 122
....

⑭ 128
- 127
....

⑮ 146
- 134
....

⑯ 452
- 175
....

⑰ 936
- 534
....

⑱ 932
- 251
....

⑲ 839
- 192
....

⑳ 417
- 166
....

1)　　867
　　 - 693

2)　　191
　　 - 150

3)　　557
　　 - 124

4)　　756
　　 - 197

5)　　852
　　 - 630

6)　　310
　　 - 204

7)　　899
　　 - 700

8)　　163
　　 - 106

9)　　584
　　 - 423

10)　 362
　　 - 341

11)　 205
　　 - 150

12)　 883
　　 - 142

13)　 778
　　 - 687

14)　 592
　　 - 440

15)　 465
　　 - 464

16)　 562
　　 - 162

17)　 669
　　 - 196

18)　 418
　　 - 101

19)　 935
　　 - 801

20)　 121
　　 - 108

① 555
  - 117

② 766
  - 658

③ 272
  - 117

④ 875
  - 659

⑤ 457
  - 124

⑥ 248
  - 203

⑦ 678
  - 625

⑧ 573
  - 478

⑨ 779
  - 437

⑩ 360
  - 169

⑪ 306
  - 230

⑫ 571
  - 283

⑬ 647
  - 380

⑭ 763
  - 140

⑮ 191
  - 189

⑯ 190
  - 179

⑰ 140
  - 115

⑱ 889
  - 381

⑲ 619
  - 558

⑳ 265
  - 210

1) 425
  - 361

2) 658
  - 336

3) 970
  - 347

4) 688
  - 254

5) 127
  - 105

6) 299
  - 274

7) 382
  - 276

8) 677
  - 281

9) 204
  - 144

10) 581
  - 542

11) 239
  - 135

12) 112
  - 110

13) 416
  - 319

14) 210
  - 157

15) 969
  - 104

16) 858
  - 271

17) 686
  - 396

18) 373
  - 325

19) 366
  - 125

20) 220
  - 168

(1) 245
   - 120

(2) 366
   - 107

(3) 393
   - 392

(4) 298
   - 260

(5) 387
   - 335

(6) 116
   - 100

(7) 360
   - 351

(8) 893
   - 146

(9) 810
   - 197

(10) 805
    - 524

(11) 639
    - 632

(12) 665
    - 589

(13) 322
    - 267

(14) 467
    - 225

(15) 982
    - 269

(16) 131
    - 116

(17) 121
    - 118

(18) 467
    - 429

(19) 820
    - 412

(20) 595
    - 126

1)　628
　 - 564
　　.............

2)　844
　 - 196
　　.............

3)　804
　 - 459
　　.............

4)　997
　 - 486
　　.............

5)　601
　 - 593
　　.............

6)　332
　 - 183
　　.............

7)　626
　 - 485
　　.............

8)　165
　 - 120
　　.............

9)　253
　 - 185
　　.............

10)　803
　 - 459
　　.............

11)　956
　 - 577
　　.............

12)　921
　 - 523
　　.............

13)　880
　 - 399
　　.............

14)　554
　 - 351
　　.............

15)　266
　 - 139
　　.............

16)　448
　 - 229
　　.............

17)　690
　 - 677
　　.............

18)　518
　 - 372
　　.............

19)　512
　 - 383
　　.............

20)　358
　 - 248
　　.............

① 104
 - 102

② 321
 - 172

③ 637
 - 577

④ 729
 - 482

⑤ 340
 - 201

⑥ 870
 - 472

⑦ 345
 - 151

⑧ 795
 - 505

⑨ 337
 - 211

⑩ 376
 - 349

⑪ 384
 - 167

⑫ 241
 - 216

⑬ 305
 - 189

⑭ 748
 - 352

⑮ 421
 - 240

⑯ 715
 - 190

⑰ 484
 - 440

⑱ 612
 - 251

⑲ 211
 - 208

⑳ 860
 - 344

(1)	250 - 147	(2)	325 - 287	(3)	763 - 572	(4)	150 - 119
(5)	673 - 375	(6)	512 - 221	(7)	661 - 653	(8)	670 - 588
(9)	308 - 138	(10)	390 - 305	(11)	780 - 290	(12)	905 - 357
(13)	701 - 213	(14)	226 - 109	(15)	821 - 676	(16)	106 - 104
(17)	884 - 636	(18)	239 - 233	(19)	288 - 132	(20)	314 - 269

① 613 − 607

② 625 − 540

③ 525 − 421

④ 479 − 456

⑤ 848 − 607

⑥ 683 − 135

⑦ 947 − 550

⑧ 316 − 200

⑨ 528 − 477

⑩ 764 − 230

⑪ 383 − 134

⑫ 407 − 276

⑬ 333 − 309

⑭ 644 − 363

⑮ 349 − 167

⑯ 586 − 234

⑰ 107 − 105

⑱ 298 − 195

⑲ 738 − 716

⑳ 587 − 501

(1)　235
　 - 227

(2)　697
　 - 437

(3)　223
　 - 146

(4)　146
　 - 112

(5)　380
　 - 313

(6)　776
　 - 446

(7)　405
　 - 379

(8)　231
　 - 150

(9)　940
　 - 346

(10)　994
　 - 797

(11)　760
　 - 126

(12)　400
　 - 200

(13)　871
　 - 104

(14)　175
　 - 103

(15)　506
　 - 413

(16)　849
　 - 625

(17)　447
　 - 436

(18)　706
　 - 168

(19)　342
　 - 251

(20)　919
　 - 271

1)　359
　　- 103

2)　822
　　- 491

3)　935
　　- 143

4)　213
　　- 142

5)　316
　　- 107

6)　202
　　- 132

7)　443
　　- 370

8)　416
　　- 360

9)　806
　　- 768

10)　227
　　- 222

11)　828
　　- 686

12)　297
　　- 218

13)　329
　　- 210

14)　716
　　- 461

15)　216
　　- 165

16)　556
　　- 528

17)　924
　　- 743

18)　928
　　- 446

19)　196
　　- 168

20)　512
　　- 158

1) 429
  - 278

2) 317
  - 247

3) 574
  - 183

4) 454
  - 313

5) 878
  - 826

6) 744
  - 147

7) 539
  - 101

8) 269
  - 266

9) 790
  - 398

10) 870
  - 139

11) 558
  - 309

12) 224
  - 114

13) 315
  - 282

14) 773
  - 215

15) 512
  - 281

16) 664
  - 609

17) 482
  - 103

18) 795
  - 262

19) 625
  - 290

20) 335
  - 233

Emma. School

THANK YOU
FOR SUPPORTING SMALL BUSINESSES LIKE US IF YOU HAVE SOME TIME, YOU CAN VISIT OUR STORE ON AMAZON, YOU MAY FIND SOME BOOKS THAT WILL HELP YOUR CHILD IN STUDYING

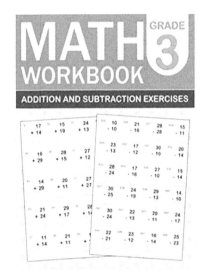
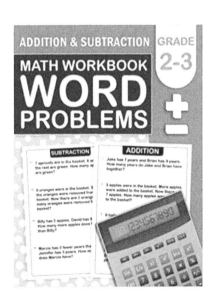

Made in the USA
Columbia, SC
25 June 2022

62226974R00030